# What Can I Be?

## STEM Careers from A to Z

TIFFANI TEACHEY

Written by Tiffani Teachey

Illustrated by Naday Meldova



ISBN: 978-0-578-62258-3

Thrive Edge Publishing

*Dedicated to my mother, **Annie Ruth Teachey**,*

*and in loving memory of my father, **Bobby Teachey I**.*

*They instilled in me the value of an education*

*and taught me if you can believe it, then you can achieve it!*

# Meet the STEM Crew

Science, Technology, Engineering, and Math (STEM) are a part of every aspect of our lives. Let's explore and take a journey with the STEM Crew kids as they represent various STEM careers from A to Z.

## Astronaut

Astronauts are trained to travel into outer space.

# B

## Biologists

Biologists are research scientists who study humans, animals, and plants.

# Civil Engineer

Civil engineers design and build roads, buildings, airports, and bridges while protecting our environment.

## Doctor

Doctors practice medicine to maintain or restore human health.

## Electrical Engineer

Electrical engineers test and develop electronics, electrical wires and electrical poles.

# Forester

Foresters manage the lands of forests, parks, rangelands, and other natural resources.

## Geologist

Geologists investigate rocks and the natural processes associated with rocks.

# H

## Hydrologists

Hydrologists study the movement of water.

# I

## Information Technologist

Information technologists develop and implement computer hardware and software systems.

# J

## Jet Engineer

Jet engineers design and build jet engines.

# K

## Kinesiologist

Kinesiologists research and analyze how the human body moves to help improve one’s health and wellness.

# Landscape Architect

Landscape architects design and beautify outdoor areas.

# M

## Mechanical Engineer

Mechanical engineers design, produce, and operate power-producing machinery.

# N

## Nurse

Nurses care for patients’ health.

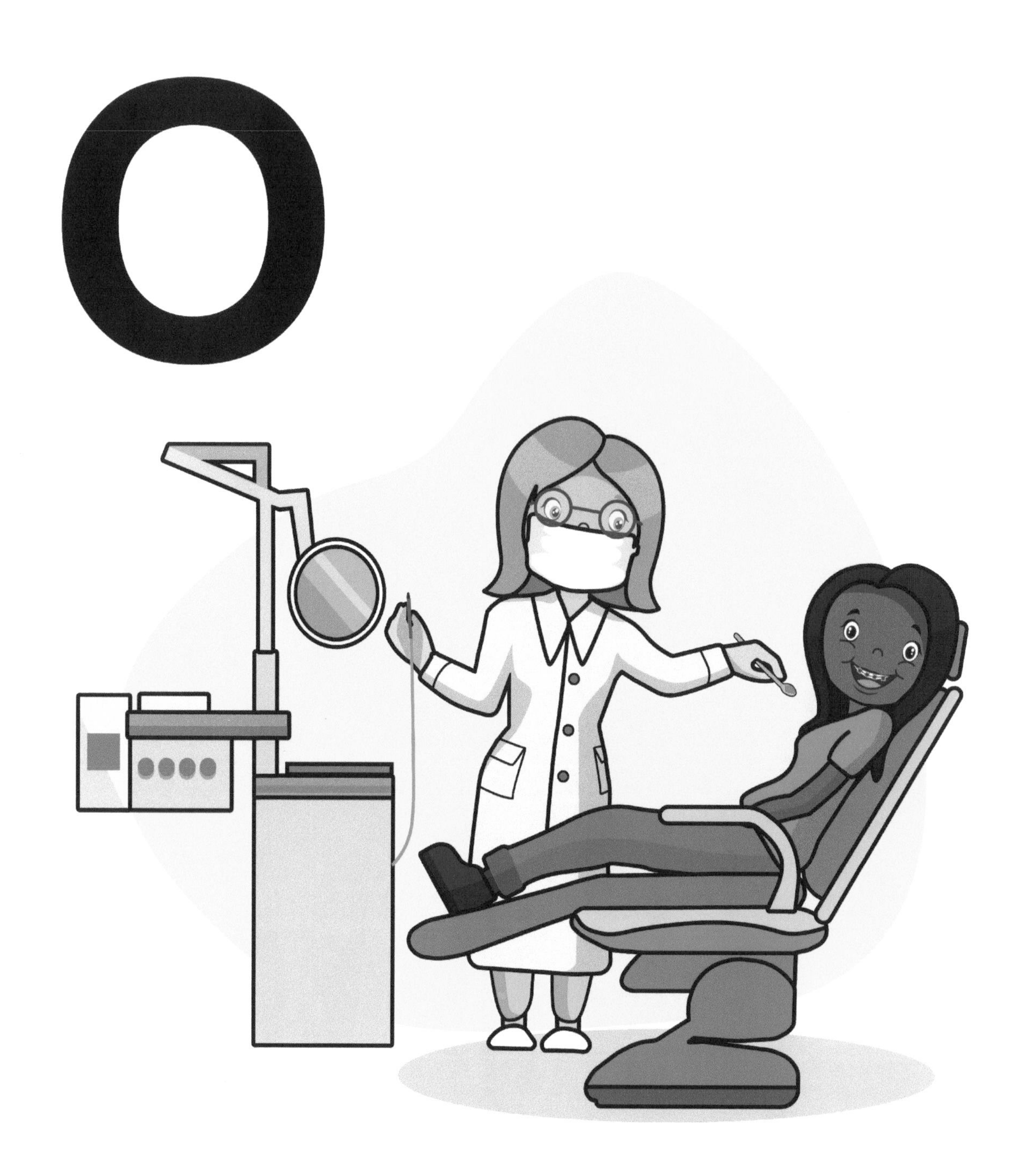

# Orthodontist

Orthodontists fix the patient's smile by using braces, retainers, and bands.

# P

## Pediatrician

Pediatricians provide medical care for infants, kids and teenagers.

## Quality Engineer

Quality engineers make sure items are of good quality.

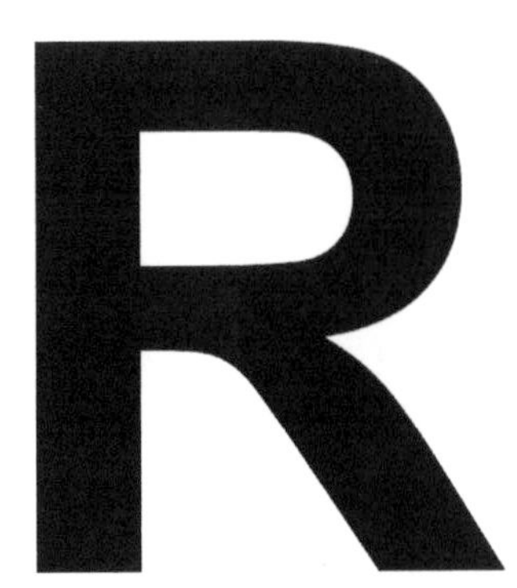

## Robotics Engineer

Robotics engineers design and build robots that do a variety of tasks.

# S

## Statistician

Statisticians collect and analyze data to make decisions.

## Transmission Engineer

Transmission engineers analyze and design electrical transmission lines and cables for power.

# Utilities Engineer

Utilities engineers design, implement, and maintain utility infrastructures.

# Veterinarian

Veterinarians diagnose, treat, research and provide healthcare for animals.

## Web Developer

Web developers build websites through writing code and working with software applications.

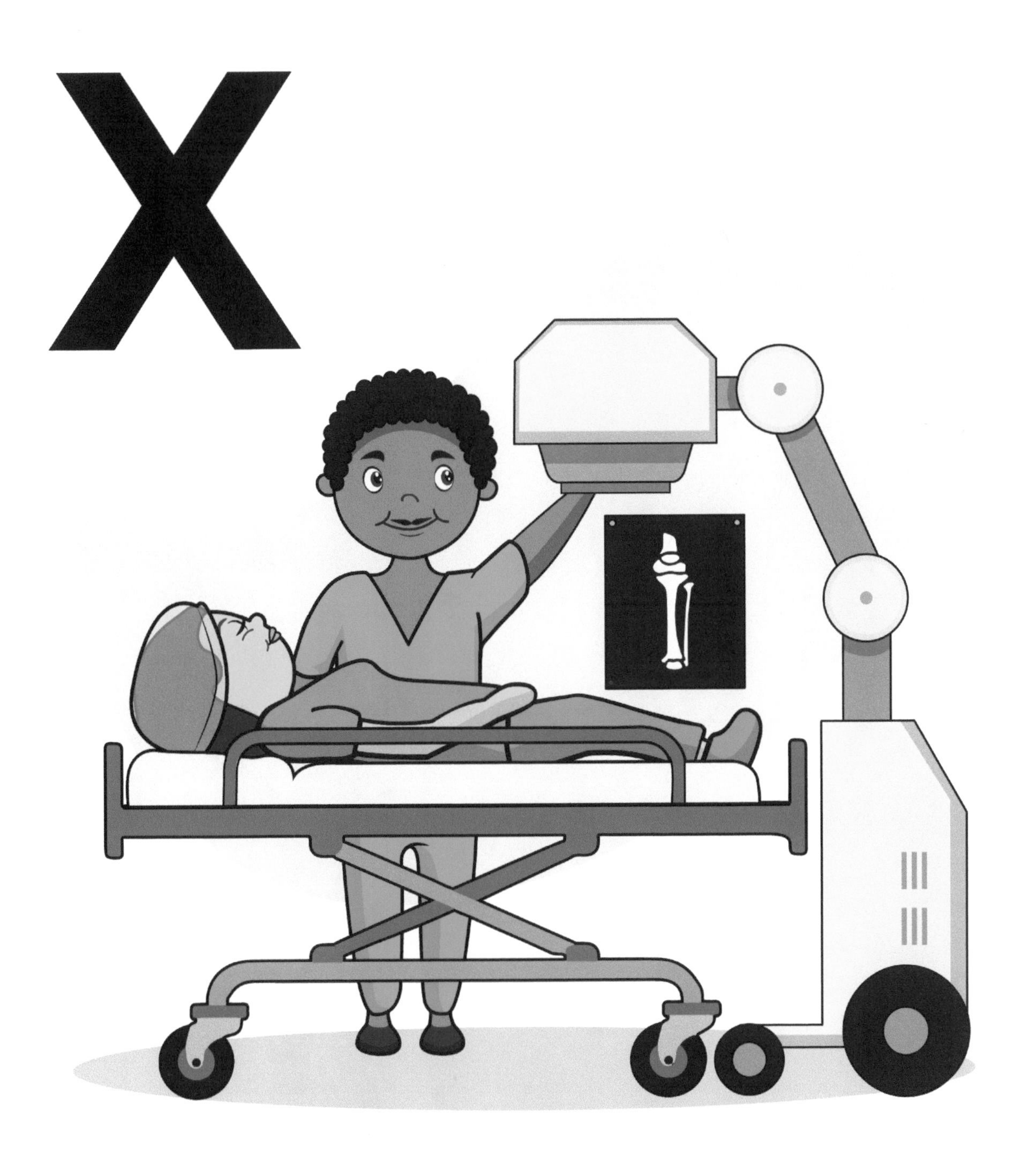

# X-Ray Technician

X-ray technicians support healthcare professionals by using x-ray equipment to take pictures of bones.

# Yacht Designer

Yacht designers design large luxury watercrafts used for leisure activities on big bodies of water like oceans.

# Z

## Zoologist

Zoologists are scientists who study animal behavior and habits.

CPSIA information can be obtained
at www.ICGtesting.com
Printed in the USA
BVHW020507121121
621380BV00007B/429